DAMAGE NOTED 3/2/94 Never

4-6

J551.2
  T   Tilling
            Born of fire : volcanoes
     and igneous rocks

DEMCO

BORN OF FIRE
Volcanoes and
Igneous Rocks

Robert I. Tilling

—an Earth Processes book—

ENSLOW PUBLISHERS, INC.

Bloy St. & Ramsey Ave.	P.O. Box 38
Box 777	Aldershot
Hillside, N.J. 07205	Hants GU12 6BP
U.S.A.	U.K.

11/91

15.95

Southeastern
Bk. Co.

Library of Congress Cataloging-in-Publication Data

Tilling, Robert I.
    Born of fire : volcanoes and igneous rocks / Robert I. Tilling.
        p. cm. — (an Earth Processes book)
    Summary: Discusses different kinds of volcanoes, their causes,
and the various igneous rocks which they produce.
    ISBN 0-89490-151-6
    1. Volcanoes—Juvenile literature. 2. Rocks, Igneous—
Juvenile literature. [1. Volcanoes. 2. Rocks, Igneous.] I. Title.
II. Series.
QE522.T55 1991
551.2'1—dc20   89—25781
                                                       CIP
                                                        AC

Printed in the United States of America

10 9 8 7 6 5 4 3 2 1

Illustration credits:
J.D. Griggs, U.S. Geological Survey, pp. 31, 35, 44, 51; Rosalind T. Helz, U.S.
Geological Survey, p. 25; Richard J. Janda, U.S. Geological Survey p. 10; Robert
Krimmel, U.S. Geological Survey, p. 4; Thomas M. Miller, pp. 16, 17, 19, 23, 42,
46, 50, 52, 54, 56; Eadweard Muybridge, p. 47; NASA/Jet Propulsion Laboratory,
p. 20; National Park Service, p. 27; Donald A. Swanson, U.S. Geological Survey,
p. 37; Robert I. Tilling, pp. 7, 9, 23, 28, 33; Lyn Topinka, U.S. Geological Survey,
p.23.

Cover photo:
J.D. Griggs, U.S. Geological Survey

CONTENTS

A 15-mile (24-km)-high plume of ash and gases rises from Mount St. Helens Volcano, during its catastrophic eruption on May 18, 1980.

1

Eruptions Can Be Hazardous

In A.D. 79 Vesuvius, which is about 9 miles (15 km) east of Naples, Italy, blew its top. Before that outburst, the volcano had been quiet for centuries, and the local people thought that it was extinct. They were wrong. Red-hot volcanic fragments and falling ash devastated the surrounding countryside, burying two wealthy Roman cities, Pompeii and Herculaneum. Thousands of people were killed. More recently, and closer to home, Mount St. Helens, in southwestern Washington, roared back to life in 1980 after sleeping for more than a century. People living in the area, just like the unfortunate citizens of Pompeii, had long forgotten that they had an active volcano for a neighbor. Following nearly two months of weak activity, an eruption on May 18, 1980 killed fifty-seven people and caused about a billion dollars worth of damage. It was the worst volcanic disaster in U.S. history.

Worldwide, more than 1,300 volcanoes have erupted at least once during the past 10,000 years. About half of these have erupted during historical times. A volcano is defined as *active* if it has erupted historically, *dormant* if it is quiet now but expected to erupt again, and *extinct* if it is not expected to erupt again. These definitions are poor, however, because the length of recorded history differs greatly from country to country. Old World countries, such as China, Italy, and

Japan, have long written histories. Their records of historical eruptions go back for thousands of years. Young countries, such as the U.S. and other countries in the New World, have written histories covering only a few centuries. The trouble is that active volcanoes can remain quiet for hundreds or even thousands of years between eruptions—much longer than human memory and many written histories. Not surprisingly, therefore, some of the worst disasters, like that at Vesuvius in A.D. 79, have happened at volcanoes thought to be extinct.

Volcanic Hazards, Risks, and Disasters

Compared to most other natural hazards—floods, storms, earthquakes, landslides—volcanic hazards occur infrequently. Yet, during the last 500 years, volcanic eruptions worldwide have caused more than 270,000 deaths and tremendous damage to property, cropland, livestock, roads, bridges, etc. The following table lists some of the worst volcanic disasters since A.D.1000.

A distinction should be made between the words *hazard* and *risk*. All active or potentially active volcanoes pose some hazard, defined as the events, products, and conditions that result from eruptions. Risk, however, is based on the likelihood of loss of life and property from such hazards. Even very hazardous zones around a volcano would pose little or no risk if few people lived or worked there. On the other hand, areas with a low incidence of volcanic hazards can be considered very high-risk if they are densely populated or highly developed.

Volcanic and related hazards can be considered as direct or indirect. Direct hazards include lava flows, pyroclastic flows, hot mudflows, ash falls, and volcanic gases. Common indirect hazards include tsunamis, cold mudflows, starvation, and atmospheric effects related to eruptions. Pyroclastic flows, tsunamis, mudflows, and post-eruption starvation have caused the worst volcanic disasters in history. Most volcanic hazards destroy only small areas on or near the volcano. However, tsunamis and ash falls from large eruptions can affect much

6

The Worst Volcanic Disasters Since A.D. 1000

Volcano	Country	Year	PRIMARY CAUSE OF DEATH				
			Pyroclastic Flow	Debris Flow	Lava Flow	Post-eruption Starvation	Tsunami
Merapi	Indonesia	1006	1,000				
Kelut	Indonesia	1586		10,000			
Vesuvius	Italy	1631			18,000		
Etna	Italy	1669			10,000		
Merapi	Indonesia	1672	300				
Awu	Indonesia	1711		3,200			
Oshima	Japan	1741					1,480
Cotopaxi	Ecuador	1741		1,000			
Makian	Indonesia	1760		2,000			
Papadajan	Indonesia	1772	2,960				
Lakagigar (Laki)	Iceland	1783				9,340	
Asama	Japan	1783	1,150				
Unzen	Japan	1792					15,190
Mayon	Philippines	1814	1,200				
Tambora	Indonesia	1815	12,000			80,000	
Galunggung	Indonesia	1822		4,000			
Nevado del Ruíz	Colombia	1845		1,000			
Awu	Indonesia	1856		3,000			
Cotopaxi	Ecuador	1877		1,000			
Krakatau	Indonesia	1883					36,420
Awu	Indonesia	1892		1,530			
Soufrière	St. Vincent	1902	1,560				
Mont Pelée	Martinique	1902	29,000				
Santa María	Guatemala	1902	6,000				
Taal	Philippines	1911	1,330				
Kelut	Indonesia	1919		5,110			
Merapi	Indonesia	1951	1,300				
Lamington	Papua New Guinea	1951	2,940				
Hibok-Hibok	Philippines	1951	500				
Agung	Indonesia	1963	1,900				
Mount St. Helens	U.S.A.	1980	60				
El Chichón	Mexico	1982	>2,000				
Nevado del Ruiz	Colombia	1985		>25,000			
TOTALS			65,200	56,840	28,000	89,340	53,090

The worst volcanic disasters since the year A.D. 1000, the number of people killed (rounded off to nearest 10), and the primary cause of death.

larger areas and places far away from the volcano. Eruption-caused atmospheric effects may even change global climate for a year or so.

Direct Volcanic Hazards

Lava flows rarely threaten human life, because people can outwalk or outrun most of them, but they can cause great property damage and loss of farmlands. Even the usually gentle eruptions of Hawaiian volcanoes can be destructive. In 1926 the fishing village of Hoopuloa, on the west coast of the Big Island of Hawaii, was destroyed by lavas erupted from Mauna Loa Volcano. No villager was hurt or killed. The 1960 eruption of Kilauea Volcano buried the small town of Kapoho on the eastern tip of Hawaii. Kilauea began to erupt again in January 1983 and has continued into the early 1990s. The eruption has already overrun many houses.

Pyroclastic flows are hot mixtures of volcanic fragments and gases. On steep slopes, they can travel at speeds as great as 95 miles (150 km) per hour, destroying everything in their paths. Such flows produced by the 1902 eruption of Mont Pelée, on the Island of Martinique in the West Indies, devastated the city of St. Pierre and killed all but two of its 29,000 inhabitants. One of the survivors was a prisoner in a thick stone-walled cell that protected him. In March–April 1982, three explosive eruptions of El Chichón Volcano, in the State of Chiapas, southeastern Mexico, caused the worst volcanic disaster in the history of that country. All villages within a 5-mile (8-km) radius of the volcano were wiped out by glowing pyroclastic flows, killing more than 2,000 people.

Mudflows are soupy mixtures of volcanic debris and water with the consistency of wet concrete. The water in them may come either from snow and ice melted by hot volcanic material or from heavy rainfall. Depending on the amount of water content and the steepness of the slope, mudflows can move at average speeds higher than 25 miles (40 km) per hour. They can also travel great distances down valleys, as much as 60 miles (100 km) or more. The greatest danger of mudflows is their tremendous power to sweep away or bury people and man-made structures. The November 13, 1985 eruption of the Nevado del Ruiz Volcano in Colombia, South America, shows that even very small eruptions can produce large, deadly mudflows. Hot ash and pyroclastic flows melted ice and snow of this 17,669 foot-high, glacier-capped Andean volcano. The meltwater mixed with the volcanic debris to form high-speed mudflows that raced down the steep valleys of the volcano to flatten towns more than thirty miles away. Over 25,000 people, mostly the inhabitants of the city of Armero, were buried by mudflows. The Ruiz catastrophe caused the worst volcanic disaster in the history of South America and the most deaths from any eruption in the world since Mont Pelée in 1902.

The fall of ash or *tephra* (a general name for airborne volcanic debris) from a large explosive eruption may cover an area of many

Ruins of a church are the only visible remains of the village of Francisco León, showing the devastation caused by pyroclastic flows of the 1982 eruption of El Chichón Volcano, southeastern Mexico.

thousands of square miles. The tephra deposits are thickest near the volcano and become thinner downwind further away from the volcano. People caught in a heavy ash fall can be injured, or even killed, by breathing the ash-choked air. But, compared to other volcanic hazards, ash falls are not especially deadly. They can cause much more property damage and nuisance, however, because of the much larger area they affect. A principal hazard is the weight of the ash itself. Roofs can collapse under the weight of ash, especially ash deposits soaked by rain. Heavy deposition of ash can kill or harm plants and wildlife, destroy or damage crops, disrupt transportation and communications systems, clog water- and sewage-treatment systems, and cause many other problems for daily living. After the May 1980 eruption of Mount St. Helens, for example, the clean-up of ash for the city of Yakima, Washington, took ten weeks and cost more than two million dollars. Yakima is about a hundred miles downwind of the volcano and received more than an inch of ash.

Sometimes volcanic gases can be killers. Such gases (for example,

Helicopter view of the city of Armero, where more than 22,000 people were killed by destructive mudflows triggered by the relatively small November 1985 eruption of Nevado del Ruiz, Colombia, South America.

carbon dioxide), which are typically heavier than air, can collect in low places to form deadly pockets. Such gas pockets are dangerous traps that cannot be seen or smelled. In February 1979, 142 people were killed by walking unknowingly into such a lethal gas pocket in the Dieng Plateau volcanic field in Java, Indonesia. Luckily, such occurrences are very rare. A much greater hazard of volcanic gases arises when they accompany and propel pyroclastic flows. Volcanic gases released during very large eruptions can also pose serious indirect volcanic hazards.

Indirect Volcanic Hazards

Tsunamis and post-eruption starvation are the worst indirect volcanic hazards. A tsunami is an extremely destructive sea wave that is also commonly, but incorrectly, called a tidal wave. Tsunamis (meaning "harbor waves" in Japanese) have nothing to do with tides. Rather, they are caused by sudden movements of the ocean floor, usually triggered by large earthquakes, but also sometimes by eruptions. A tsunami generally occurs as a series of waves, rather than as a single one. In deep ocean water, such waves may travel at speeds greater than 300 miles (500 km) per hour. In deep water, the height of these waves is no more than three feet (1 m), harmless and not noticed. But as the tsunami approaches shallow coastal waters, huge waves as high as 170 feet (50 m) can form as they "touch bottom."

Tsunamis can devastate low-lying coastal areas, even those far from the triggering eruption or earthquake. Perhaps one of the best examples of the hazards of an eruption-caused tsunami is the 1883 eruption of Krakatau Volcano, in the Sunda Straits of Indonesia. Close to the eruption, tsunamis washed away 165 coastal villages on the Indonesian islands of Java and Sumatra, killing 36,000 people. The tsunami was detected by the tide gauge in the harbor of Aden, on the southern coast of the Arabian Peninsula, more than 4,350 miles (7,000 km) from Krakatau.

The April 1815 eruption of Tambora Volcano in the Lesser Sunda

Islands, Indonesia, the largest eruption in the world in historical times, was also the most deadly. It killed an estimated 92,000 people, about 12,000 by the eruption itself and the rest by post-eruption starvation. The thick blanket of volcanic debris deposited and the huge amount of gas released combined to destroy all croplands as well as most animal and plant life. As a consequence, the people living in the region had little or no food, and died of disease and starvation in the months following the eruption.

The fine ash and gases that the Tambora eruption injected into the atmosphere formed a towering volcanic cloud, which drifted many times around the world. This cloud was dense enough to screen out some of the sun's heat, and to lower average global temperatures, in places as much as 4-6° F (2-3° C), but more commonly about one degree. The volcanic dust in the atmosphere produced unusually brilliant and colorful sunrises and sunsets for many months. For more than a year after the eruption, record cold temperatures, unusual snows and frosts occurred during the normally hot summer months in many countries in the northern hemisphere. The global climate change was so dramatic that 1816 was called "the year without a summer."

In June 1783, the largest eruption in Iceland's history took place at Laki Volcano. Though not especially explosive, this eruption poured out a tremendous volume of lava—more than three cubic miles (12 cu. km)—over a period of eight months. The erupted lava was also unusually high in a poisonous gas, sulfur dioxide. The gas cloud produced a bluish haze that ruined the summer crops and killed about 75 percent of all livestock in Iceland (11,000 cattle, 28,000 horses, and 190,000 sheep). Within the next few years, eruption-caused starvation resulted in the death of about 10,000 people, or about a quarter of Iceland's total population at the time.

The volcanic haze from the Laki eruption drifted eastward, first reaching Europe and then, after about two months, China. The Europeans were puzzled by the occurrence of this strange haze and called it a "dry summer fog." The unusual summer was then followed by an exceptionally cold winter in Europe. This weird weather pattern impressed Benjamin Franklin, who was at the time the U.S. Ambassador to France and living in Paris. He reasoned that the "vast quantity of smoke" from the Laki eruption was causing the "universal fog" as well as the severely cold 1783–84 winter. Franklin was the first person to figure out that volcanic eruptions could affect climate. He wrote a scientific paper on his idea in May 1784, which he presented in December 1784 at a scientific meeting in Manchester, England. Modern scientific studies of the atmospheric effects of large eruptions have shown Ben Franklin's brilliant idea to be correct.

Scientific Studies Can Reduce Volcanic Hazards

From earliest history, volcanic eruptions have both fascinated and terrified mankind. They still do. Ancient civilizations believed that eruptions were caused by angry gods, goddesses, or other supernatural beings or powers. We now know that eruptions are natural happenings and can be studied scientifically. The following chapters will show that scientists are beginning to understand how volcanoes actually work. A good knowledge of volcanic behavior—before, during, and after eruptions—is the starting point for any future reduction in the loss of life and property from volcanic hazards.

2

Volcanoes are Found in Special Places

What exactly is a volcano? It is a mountain or hill built by the piling up of materials erupted from openings in the earth's surface. The term *volcano* is also used for the opening (or vent) itself. The word *volcano* comes from the Italian volcanic island of Vulcano. Centuries ago, Vulcano was believed to be the chimney of the forge of Vulcan, the blacksmith of the Roman gods. Vulcano, which last erupted in 1886, is still considered an active volcano. A volcanic vent can be a single hole, several connected holes, or a crack or fissure. Most volcanoes contain only one vent, generally located at the summit, but some have one or more vents on their sides (or flanks) in addition to the one on top. On a few volcanoes, many vents are strung out in long narrow bands called *rifts* or rift zones.

There are more than six hundred active volcanoes on land; many more lie unseen beneath the sea. In recent years, scientists in research submarines have discovered and mapped several active volcanic areas on the ocean floor. The earth's active volcanoes are not found everywhere, but only in certain regions. Most of the above-sea volcanoes are found along, or near, the edges of the continents; many volcanoes are islands. The United States has about sixty active volcanoes. Except for those in Hawaii, all are located in the states of the western (Pacific) edge of the North American continent: California, Oregon, Washington, and Alaska. Similarly, nearly all of the active volcanoes of Central and South America are found along the western (Pacific) border of the continent.

The Circum-Pacific Ring of Fire

More than half of the earth's above-sea volcanoes are found around the Pacific Ocean. They form a notorious zone called by some the "Ring of Fire." Large, often destructive earthquakes also occur in this same zone. The volcanoes of the Cascade Range of the U.S. Pacific Northwest and Canada, and those along the Alaskan Peninsula and on the Aleutian Islands, form the North American segment of the Ring of Fire. The circum-Pacific volcanoes erupt frequently, and their eruptions have caused some of the worst volcanic disasters in history.

Plate-Tectonics Theory

The fact that volcanoes occur along the edges of some continents or form island chains was not well understood until recently. Most scientists now agree that the locations of the earth's volcanoes are best explained by the plate-tectonics theory, which became generally accepted in the 1960s. According to this theory, the outer part of the earth is broken into about a dozen rigid slabs or plates. These rigid plates are at least 50 miles (80 km) thick on average. They move relative to one another above a deeper, hotter, and more mobile zone at average speeds of a few inches a year—about the same rate that human fingernails grow. The geologic forces and processes resulting from plate motion are concentrated in the narrow boundary zones between the moving plates. Geologists recognize three common types of plate boundaries: divergent, convergent, and transform-fault.

Divergent (or spreading) boundaries separate neighboring plates that are pulling apart. Perhaps the best example of this type of boundary is the Mid-Atlantic Ridge. It separates the North and South American plates from the Eurasian and African ones. As the plates pull apart, lava erupted from fissures along or near the boundary fills the gap by adding new material to the plates. Almost all of the spreading boundaries, and the active volcanoes along them, are covered by the oceans. They can be mapped and studied only by use of ship-towed, remotely controlled photographic and other scientific instruments and, more recently, by scientists in deep-diving submarines. A notable exception is Iceland, where the Mid-Atlantic Ridge is exposed on land. Iceland is entirely formed of volcanic rocks and has many active volcanoes.

Convergent boundaries, also called subduction zones, characterize places where plates collide, and one of them is dragged down (subducted) beneath the other. Good examples of convergent boundaries are the Aleutian Trench, where the Pacific Plate is being subducted

ACTIVE VOLCANOES OF THE WORLD

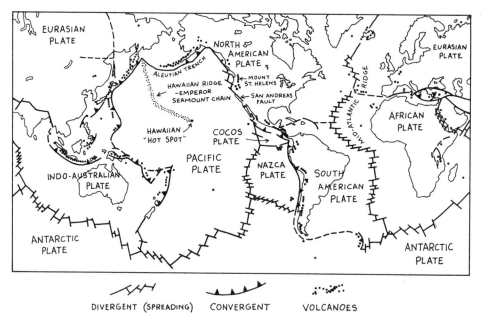

DIVERGENT (SPREADING) CONVERGENT VOLCANOES

Map showing the principal tectonic plates and the location of some of the nearly 600 active volcanoes on earth.

under the North American Plate, and the Peru-Chile Trench, where the Nazca Plate is being subducted beneath the South American Plate. As subduction continues, one plate is pulled deep into the earth, where temperatures are high enough to partly melt solid rock. This newly formed molten rock (magma) is lighter than the surrounding solid rock. It rises along cracks and weak zones and erupts, forming new volcanoes or feeding existing ones. Thus a rough balance is achieved, as the material melted from one plate during subduction is ultimately erupted and added to the other plate. Most convergent-plate boundaries are found next to the edges of some continents or certain island chains. This fact explains why most of the world's active volcanoes are in the circum-Pacific region.

A *transform-fault* plate boundary is one where one plate slides horizontally past another. Such boundary zones are better known for

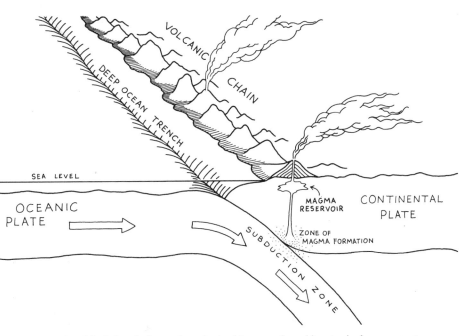

This simplified sketch shows the principal features found in a typical convergent-plate boundary between an oceanic and a continental plate moving toward one another.

having many earthquakes rather than containing many active volcanoes. The best-known and most-studied example is the San Andreas Fault zone of California, which separates the Pacific and North American plates.

The Hawaiian "Hot Spot"

Most of the world's active volcanoes are located along or near the boundaries of the earth's moving tectonic plates. Some, however, are found in the middle of some plates and are called *intraplate* volcanoes. A prime example are the volcanoes that form the Hawaiian Islands. The Hawaiian volcanoes are located nearly 1,900 miles (3,000 km) from the nearest boundary of the Pacific Plate. The growth of the Hawaiian and, possibly some other intraplate volcanoes, can also be explained by the plate-tectonics theory.

The Hawaiian volcanoes are the exposed peaks of a huge, submarine volcanic mountain range called the Hawaiian Ridge-Emperor Seamount Chain, which stretches from the Aleutian Trench to the Hawaiian Islands. All of the volcanoes in this chain except for those in Hawaii are now extinct. The Hawaiian-Emperor Chain was formed over a long time by the continuous northerly movement of the Pacific Plate over a deeper, fixed hot spot. This hot spot provides enough heat to melt partially the material below the Pacific Plate as it passes over. The new magma produced then rises and erupts onto the seafloor to form a volcano. With countless eruptions of lava, the new volcano grows larger and higher, until it finally pokes above the sea surface to form an island volcano. Eventually though, the moving Pacific Plate carries the volcano away from the hot spot, and, cut off from its feeding magma source, the volcano becomes extinct. But as one volcano dies another begins to grow over the hot spot, thus starting another cycle of volcano birth, growth, and death. Over a period of about 70 million years, this volcanic cycle was repeated many times to produce the Hawaiian-Emperor Chain.

The Kilauea and Mauna Loa volcanoes on the Island of Hawaii

(also called the Big Island), the southeasternmost and youngest island of the chain, are currently above the hot spot and tap the magma produced by it. According to the hot-spot theory, Kilauea and Mauna Loa should eventually become extinct, and a new volcano should form east or south of the Big Island. Even though Kilauea and Mauna Loa are still very much active, a new submarine volcano, called Loihi, is already growing about 20 miles (30 km) off the Big Island's south coast. The top of Loihi, however, is still more than a half mile (nearly a kilometer) below the ocean surface. If Loihi remains continuously active in the future, tens of thousands of years will still be required for it to grow high enough to form a new Hawaiian island. Or possibly, it might become extinct without reaching the surface. Time will tell.

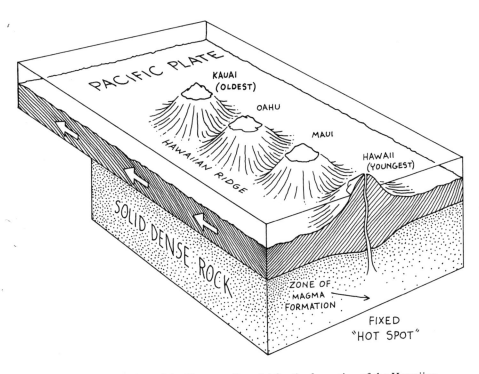

This is an illustration of the "hot-spot" model for the formation of the Hawaiian volcanic chain. (Modified from an illustration provided by Maurice Krafft, Centre de Volcanologie, Cernay, France).

The hot-spot theory seems to work for the intraplate volcanoes of Hawaii. Some scientists have called upon several other hot spots, under continents as well as oceans, to explain intraplate volcanoes elsewhere on earth. However, much more data will be needed to test the theory worldwide. At present, how most intraplate volcanoes work is not well understood.

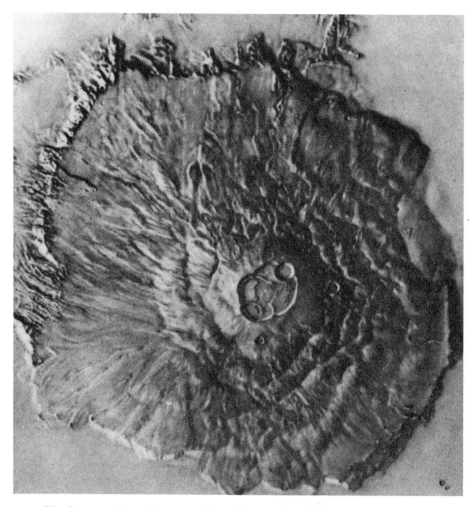

The Olympus Mons Volcano on Mars is larger than the entire Hawaiian island chain.

Extraterrestrial Volcanoes

Volcanoes are found not only on our planet. Since the space age began, planetary explorations, manned and unmanned, by United States and Soviet Union space agencies have found dramatic evidence of extraterrestrial volcanoes. Volcanic eruptions have taken place on the Moon, Mars, Venus, (possibly) Mercury, and on several smaller bodies. One of the most important discoveries has been that extraterrestrial volcanoes and volcanic features can be much larger than those on Earth. For example, the Martian volcano Olympus Mons is 17 miles (27 km) high and 350 miles (560 km) across. Another exciting, but lucky, discovery involved Voyager 2 spacecraft images made in 1979 of Io, one of the moons of the planet Jupiter. These live images captured some volcanic eruption plumes (of sulfur dioxide gas) more than 90 miles (150 km) high. This finding was very exciting, because it was the first direct evidence that eruptions could be taking place right now in other parts of our solar system. Ten years later, images of Neptune's moon Triton, sent by the still-working Voyager 2 spacecraft, showed dark, eruption-like plumes (of nitrogen ice and gas) trailing about 90 miles (150 km) downwind from volcanic vents. In contrast to this recent volcanic activity on Io and Triton, volcanism on the Moon, Mars, Venus, and Mercury occurred only very early in their histories, several billion years ago.

So far, scientists have not found any convincing evidence of plate-tectonics movements on our neighbor planets. Volcanic activity on Earth is clearly linked to plate-tectonics processes, but the forces that cause extraterrestrial volcanism are not understood.

3

Types of Volcanoes and Eruptions

Volcanoes come in different sizes and shapes. These differences are related to the way volcanoes erupt, which in turn depends on the kind of magma erupted. *Magma* is what scientists call molten rock while it is underground. Once erupted, magma is called *lava*, which can take many forms and is given special names.

Types of Volcanoes

There are four main types of volcanoes: composite volcanoes (also called stratovolcanoes), shield volcanoes, cinder cones, and lava domes. *Composite* volcanoes, the most abundant type, form steep, cone-shaped mountains. Mount St. Helens is a composite volcano, as is Mount Fuji in Japan. Such volcanoes are formed by layers of ash, other volcanic debris, and lava flows. In contrast, *shield* volcanoes are gently-sloping volcanic mountains built almost entirely of lava flows. They were so named on account of their shape, which is reminiscent of an ancient warrior's shield. Mauna Loa on the Island of Hawaii is the world's best—and largest—example of a shield volcano. These volcanoes are typically much larger than composite volcanoes.

Cinder cones and *lava domes* are generally much smaller than

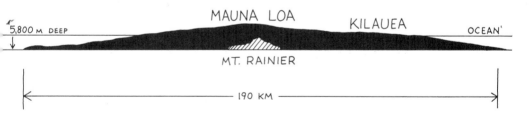

Comparison of volcano types (clockwise starting at top left). 1. Lava dome—growing inside the large crater at Mount St. Helens; 2. Shield volcano—Mauna Loa on the Island of Hawaii, viewed from the Hawaiian Volcano Observatory; 3. Mauna Loa and Mount Rainier volcanoes drawn at the same scale show that shield volcanoes are generally much larger than composite volcanoes.

)

shield or composite volcanoes. Cinder cones are formed by the accumulation of small lava fragments (cinders) blown out from a single vent. These steep-sided cones usually have a small depression or crater on top. Lava domes (also called volcanic domes) are roundish mounds formed by the piling up of lava around a central vent. Lava domes commonly form inside the summit craters of composite volcanoes, as at Mount St. Helens.

Types of Eruptions

Worldwide, about fifty to sixty volcanoes erupt each year on average. This average rate of eruptions has probably not changed for centuries. Practically all of the reported eruptions are of volcanoes on land, because eruptions of submarine volcanoes, with rare exceptions, are not detectable.

What causes a volcano to erupt? The ancient Romans believed that the hot materials thrown from volcanoes came from the forge of Vulcan, the blacksmith of the Roman gods, whenever he made thunderbolts for Jupiter, king of the gods, and weapons for Mars, the god of war. Hawaiian and other Polynesian legends put the blame on Pele, the goddess of volcanoes. Pele could cause eruptions and fires simply by digging her magic stick into the ground. Other peoples and cultures also linked volcanic outbursts to supernatural individuals or events. Scientists now know that eruptions are caused by the escape of gases contained in magma.

When first formed, all magmas consist mainly of liquid, along with a small portion of solid matter. The liquid is molten rock with gases dissolved in it; the solids are suspended minerals and, possibly, fragments of surrounding rock. The dissolved gases are mostly water vapor, together with much smaller amounts of gaseous compounds of carbon, sulfur, chlorine, and fluorine. These gases remain dissolved in the liquid under the high pressures and temperatures deep in the earth. But as the magma rises, the pressure and temperature acting on it decrease. More minerals crystallize from the liquid, and the gases become more concentrated in the liquid that remains. As the process continues, these two effects combine to start the separation of the dissolved gases from the liquid. From this point on to eruption, the magma is made up of three distinct parts: liquid, solid, and gas bubbles.

As the magma continues upward, more gas bubbles form and the existing bubbles expand. This process of gas separation and expansion increases the internal or gas pressure of the magma. When the magma pressure becomes greater than the strength of the surrounding rocks, the magma can break through to erupt. It is the release of the magma's pent-up gas pressure that triggers eruptions.

The rate of the gas release determines whether the eruption is violently explosive or relatively gentle. If the magma (lava) is highly

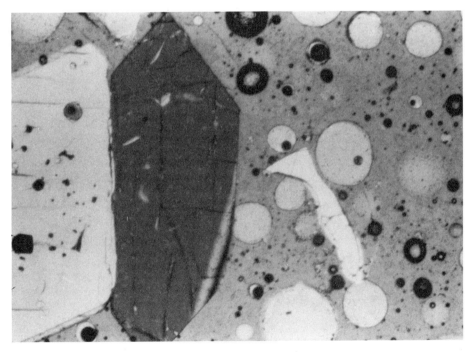

Microscope view (0.1 inch [2 mm] wide) of Hawaiian lava showing the three parts that make up magma: original liquid rock that cooled rapidly to form natural glass; solids, here as olivine crystals (large pieces to left); and gas trapped when lava solidified, as shown by gas-bubble holes called vesicles.

fluid, as in the case of Hawaiian volcanoes, the expanding gases can escape easily. The resulting eruption is nonexplosive or weakly explosive. However, by definition, any eruptive process must be somewhat explosive. Even though Hawaiian eruptions are considered nonexplosive, they can shoot lava more than 1,800 feet (550 m) into the air. If, on the other hand, the lava is thick and pasty, volcanic gases cannot escape easily. The expanding gases can then build up very high pressures and ultimately escape with explosive energy upon eruption. The explosive force rips the erupting lava into tiny bits, sending up a plume of ash and gases many thousands of feet high.

An explosive volcanic eruption can be compared with the sudden release of gas from soda pop or the uncorking of a bottle of champagne, especially if the beverage has been shaken up. Shaking of the beverage separates more gas from the liquid and forms more gas bubbles, which expand suddenly (explode) when the thumb or cork is abruptly removed.

Most composite volcanoes, such as Mount St. Helens and other subduction-zone volcanoes, tend to erupt explosively. In contrast, intraplate volcanoes, such as the Hawaiian shield volcanoes, generally erupt nonexplosively. The more fluid and hotter the lava, the smaller the chance for explosive eruptions. Fluid lavas tend to contain more iron (Fe), magnesium (Mg), calcium (Ca), and titanium (Ti), and less silicon (Si), aluminum (Al), sodium (Na), and potassium (K) than sticky, stiff lavas. The table on page 28 summarizes the general associations between volcano type, lava type, and eruption style.

Volcanoes can erupt in many ways and, like people, display distinctive individual behavior. Nonetheless, scientists have recognized several general types of eruptions commonly observed at many volcanoes. These eruption types are named for some volcano or volcanic region that well illustrates a particular kind of eruptive activity:

Hawaiian—nonexplosive eruptions of very fluid lava, mainly in the form of lava flows and lava fountains, typical of those in Hawaii.

Strombolian—named for Stromboli Volcano, Italy; typically many small, weakly explosive eruptions in fairly regular short bursts.

A lava fountain, 1,800-feet (550-m) high, during the 1959 Kilauea Iki eruption of Kilauea Volcano, Hawaii. Though impressive, this fountain is small as compared with ash plumes produced by some explosive eruptions of composite volcanoes (compare with 15-mile [24-km]-high plume of Mount St. Helens on May 18, 1980 found on page 4 of this book).

Vulcanian—named for Vulcano Volcano; similar to Strombolian activity, but more explosive; commonly produces plumes of ash, but few lava flows.

Peléean—named for Mont Pelée; commonly involves the eruption of viscous lava to form volcanic domes. Violent explosions or collapses of these domes can trigger high-speed deadly pyroclastic flows.

Plinian—named after Pliny the Elder, the famous Roman naval officer and scholar, who was killed while observing the A.D. 79 eruption of Vesuvius that buried Pompeii. These are the most powerful eruptions known, involving large volumes of viscous lava. Large Plinian eruptions can produce ash plumes tens of miles high. The ash fall from these plumes can affect wide areas hundreds of miles downwind.

Phreatic—a term derived from the Greek word for *well*; generally minor volcanic activity driven by explosively expanding steam resulting from cold groundwater coming into contact with hot rock. These steam-blast explosions only erupt fragmented preexisting solid volcanic or other rocks; no new magma is erupted.

Volcanoes, Lava, and Eruptions

Volcano Type		Lava Type		Eruption Style
Larger	**smaller**	**viscosity ("stiffness")**	**magma composition**	
Shield	cinder cone	Very fluid	Basaltic (High in Fe, Mg, Ca, and Ti)	Generally non-explosive but sometimes weakly explosive
	cinder cone	Fluid to viscous	Andesitic	Generally explosive but sometimes non-explosive
Composite	lava dome	Very viscous	Dacitic to Rhyolitic (High in Si, Al, Na, and K)	Typically highly explosive but can be non-explosive, especially after a previous large explosion

General associations between volcano type, magma composition, lava viscosity, and eruption style. There are exceptions and variations between these common types and styles.

The size of an eruption is rated by the Volcanic Explosivity Index (VEI), which is based on the amount of lava erupted, the height of the ash plume, and other measures of the energy released. The VEI scale can range from 0 to 8 or even higher. A rating of 0 is for a nonexplosive eruption, a rating of 8 for a highly explosive one. For example, the eruption of Tambora Volcano in 1815, the world's largest historical eruption, is given a VEI of only 7. The 1980 eruption of Mount St. Helens barely rates a 5. Explosive eruptions that would qualify for VEI ratings of 8 or greater all occurred thousands or millions of years ago, long before the emergence of civilizations. Examples are the gigantic eruptions that produced huge calderas in the Yellowstone, Wyoming, and Long Valley, California areas. A *caldera* is a large crater formed by ground collapse associated with eruption of large volumes of ash.

Scientists now have a good general understanding of what causes a volcano to erupt, and of what controls eruption style. For some well-studied volcanoes, it is possible to predict when they might erupt. But scientists still do not have the answer to the basic question of what causes an eruption to stop once it has started. An eruption may stop for a number of reasons: the volcano's supply of magma may have been entirely used up during the eruption; the volcano may simply have run out of steam—that is, the escaping gases causing the eruption were used up, even though more magma was available; or possibly the eruption stopped because the volcanic vent became clogged by solid debris or hardened lava. In the case of a blocked vent, however, gas pressure acting on the magma could build up again and possibly lead to another eruption, perhaps even more explosive than the first one. In rare cases, the reasons why the eruption stopped can be determined in hindsight after careful analysis of all the scientific facts. So far, however, no person or scientific study has been able to predict successfully, either before or during an eruption, how long it might last.

4

Volcanic Products and Structures

Volcanic products vary greatly in size and physical appearance, depending on volcano type, lava composition, and eruption conditions. Most volcanic products fall under one of two categories: lava flows—formed during nonexplosive eruptions by the cooling and hardening of flowing lava; and fragmental products—formed during explosive eruptions by the tearing apart of new liquid lava or the shattering of old solid rock. These products can accumulate in various ways to form different kinds of volcanic rocks, deposits and structures.

Lava Flows

After nonexplosive activity begins, the erupted lava soon forms one or more red-hot (incandescent) streams. These lava streams then begin to flow down the side of the volcano. Several small streams can join to form a lava river. Lava flows can either follow existing valleys and low areas or form their own paths. If the lava is fluid, the flows can travel ten or more miles from the vent. On steep slopes, fluid lava flows can move quite rapidly. Some lava flows on steep slopes in Hawaii have been clocked at speeds greater than thirty miles (50 km) per hour,

Aerial view showing part of an active 8-mile (13-km)-long lava flow from Pu'u O'o vent (top), Kilauea Volcano, Hawaii, in July 1983.

but average flow speeds of six miles (10 km) per hour or less are much more typical. If the lava is relatively viscous, the flows move sluggishly and travel only short distances from the vent.

There are three main types of lava flows. *Pahoehoe* (pronounced pah-hoy-hoy) is the name given to solidified fluid lava whose surface is smooth, gently rolling, and wrinkled to give a ropey appearance. In contrast, a type of lava called *aa* (pronounced ah-ah) has a very rough, jumbled surface composed of jagged fragments and blocks. Both of these names are Hawaiian but are now used worldwide for similar lava flows. When pahoehoe lava is erupted underwater, it commonly forms a special kind of flow called *pillow lava*, because it resembles a pile of sacks or pillows. Pillow lava makes up much of the deep ocean floor.

The third major type of lava flow, called *block* lava, forms during nonexplosive eruptions of viscous lava. Block lava is made up of large angular blocks—bigger but less jagged than those in aa lava—and forms thick flows that move only small distances from the vent. Pahoehoe lava flows generally are thin, rarely more than twenty feet (6 m) thick. Aa and block lava flows can be several times thicker. Pahoehoe and aa flows are known to occur on land and beneath the sea, but no block lava flows have yet been found on the deep ocean floor.

Fragmental Products

During explosive eruptions, liquid lava is torn into fragments and hurled into the air. Volcanic explosions can also shatter and throw out pieces of the old solid rocks around the vent. All such materials are called *pyroclastic* (''fire-broken'' in Greek) regardless of their size and shape, and whether they are derived from new liquid lava or from old solid rock. A few volcanoes only erupt explosively and are entirely built of pyroclastic materials. Most composite volcanoes, however, are made up of both fragmental products and lava flows. This indicates that such volcanoes can switch from explosive to nonexplosive activity from one eruption to the next, or even during the same eruption.

Fragmental volcanic products differ greatly in size and form, depending on the type of lava ejected and the explosive force of the eruption. In general, the more fluid the erupting lava, the weaker the explosive force. If the explosive force is weak, the fragments tend to be larger in size because they are less broken up. In contrast, violent explosions erupt viscous lava that breaks up into smaller fragments, consisting mostly of *lapilli* ("little stones" in Italian), ash, and dust. However, the sizes of the fragments in any given explosive eruption can range from room-size blocks to dust-size particles. The largest and heaviest fragments are found closest to the vent.

During weak explosive eruptions of fluid lava, the most common products are irregular fragments of bubbly (gas-rich) lava called *scoria* or *cinder*. If the scoria has a great many gas-bubble holes (vesicles), it is called *pumice*. Pumice can be light enough to float on water, if its vesicles are not interconnected. Scoria or pumice fragments, if still in a mostly liquid state when they hit the ground, can

Common Fragmental Volcanic Products

Size (average diameter)	Shape	Condition when erupted	Some Examples
Greater than about 2 1/2 inch (64 mm)	Angular	Solid	Blocks
	Round to slightly angular	Semi-solid	Bombs, cinders, scoria
		Largely liquid	Pumice, spatter
Between about 2 1/2 inch (64 mm) and 1/10 inch (2 mm)	Round to angular	Liquid or solid	Lapilli, cinders, scoria, pumice, spatter
Smaller than about 1/10 inch (2mm)	Generally angular, but sometimes round	Liquid or solid	Ash, dust (if finer than 1/16 mm)

A classification of common fragmental volcanic products.

flatten or splash; such fragments are called volcanic *spatter*. Over the centuries, people in many countries have used pumice as a lightweight but sturdy building material, or as an abrasive agent for cleaning and industrial purposes.

Fragments larger than one and one quarter inches (32 mm) are called volcanic *bombs* if erupted in a semisolid state and fluid enough to change shape while sailing through the air. Bomb-size, but angular, fragments are called *blocks*; these were thrown out in a state too solid to allow any change in shape during flight.

Fragmental Deposits

Explosive eruptions of sticky, stiff lava commonly produce one or more of these fragmental deposits: pyroclastic flows, ash falls, and volcanic mudflows. Pyroclastic flows, also called ash flows, consist of glowing, hot mixtures of pumice and ash. The solid fragments are carried and swept along in a cloud of expanding volcanic gases. These pumice-ash-gas mixtures are heavier than air but highly mobile. They form ground-hugging flows that can sweep down steep slopes and valleys at high speeds and cover great distances. Incandescent pyroclastic flows are also called glowing avalanches. The combination of high speed, temperature, and gas content makes pyroclastic flows especially destructive and deadly.

While pyroclastic flows rarely travel more than ten miles from the vent, ash falls from a large eruption plume may extend downwind to distances of hundreds of miles from the volcano. The volcanic deposits from ash falls are thickest in the immediate area of the vent and gradually thin downwind. Because the ash fragments are much cooled (by passage through air) by the time they hit the ground, ash falls are less dangerous than pyroclastic flows.

Explosive volcanoes capped by snow and ice or located in rainy mountainous regions commonly produce volcanic mudflows during or shortly following eruptions. Mudflows are mixtures of fragmental volcanic debris and water. If saturated with water, mudflows can

Examples of some common fragmental volcanic products (coin gives scale), clockwise from upper-left corner: pumice, volcanic bombs, spatter (Pele's Tears), and lapilli.

become unstable and flow very rapidly down mountain valleys. Mudflows can be hot or cold, and very destructive in either case. Hot (or primary) mudflows are formed when the erupted materials cause melting of snow or ice and the resulting meltwater mixes with the volcanic debris. This was the case in the 1985 volcanic disaster at Nevado del Ruiz. Hot mudflows may also happen if an eruption occurs during heavy rain. Cold (or secondary) mudflows can occur on steep volcanic slopes during or following heavy rainfall, though these cases involve cold fragmental deposits from previous eruptions.

Volcanic Gases

Escaping gases provide the driving force for eruptions. The rising plumes generated during explosive eruptions are mixtures of volcanic ash, dust, and gases. Large explosions can produce eruption plumes tens of miles high, punching high into the atmosphere. The plume materials form a volcanic cloud that may last for a long time, drifting many thousands of miles and circling the globe repeatedly. Nearly all of the ash fragments drop out of the cloud within a few weeks, and even the finest dust particles within a year or so. But the volcanic gases, and droplets of acids formed from them, may remain in the atmosphere for years. For example, the volcanic cloud produced by the March–April 1982 eruption of El Chichón Volcano lasted through 1985.

Volcanic Structures

Countless eruptions of fluid, far-traveling lava flows around a central vent can build a volcanic shield (shield volcano), provided the feeding vent remains active in the same place for long enough. The large Hawaiian volcanoes are built almost entirely of fluid lava flows and have the distinctive shieldlike forms. Hawaiian and most other large volcanoes have a large depression, called a *caldera*, on their summits. Calderas are formed by collapse of the volcano's summit when the magma below the volcano is removed because of eruption. Similar but smaller collapse structures are called *craters, pit craters,* or *pits.*

The accumulation of many short lava flows, interlayered with ash and other fragmental deposits, results in much steeper, cone-shaped structures called composite volcanoes or *stratovolcanoes*. If extremely stiff lava is erupted, the sluggish flows formed can barely move before hardening completely. The lava simply piles up in the immediate vicinity of the vent to form roundish mounds called volcanic domes or lava domes. The dome-forming process is similar to the slow, steady squeezing of an upward-pointed tube of toothpaste. Lava domes are commonly formed after a big explosive eruption.

Some of the largest volcanic structures in the world are not shaped like a shield or a composite volcano. Instead, they form flattop mountains, called volcanic plateaus, or extensive flat-floor valleys (volcanic plains), which can cover tens of thousands of square miles. These gigantic volcanic features can be formed either by explosive or by nonexplosive eruptive activity. They may be the result of repeated

The Columbia River Plateau west of Little Goose Dam, Washington, showing a high cliff exposing many basaltic lava flows. A car on highway at base of cliff (small white dot, lower left) gives scale.

high-volume but nonexplosive outpourings of fluid lava from long fissure vents. An excellent example of this manner of formation is the Columbia River Plateau in the U.S. Pacific Northwest. A volcanic plateau or plain may also result from huge explosive eruptions that produce highly mobile, gas-rich pyroclastic flows. Good examples of this type are found in the Yellowstone region of Wyoming and in New Zealand.

5

Igneous Rocks

The earth's crust is made up of three kinds of rocks: igneous, sedimentary, and metamorphic. Sedimentary rocks form on the surface and are the products of weathering and erosion of older rocks; some, such as limestone, form by the accumulation of minerals deposited from oceans or lakes. Metamorphic rocks are formed deep in the earth by the changing of older rocks into new rocks under high heat and pressure; the name *metamorphic* means "change of form" in Greek. Igneous rocks are those formed from magma; for that reason, they also are called magmatic rocks. The word *igneous*, like the words *ignite* or *ignition*, comes from *ignis*, the Latin word for "fire."

There are two main groups of igneous rocks: volcanic and plutonic. Volcanic (extrusive) rocks include all the products resulting from surface eruption of lava. Plutonic (intrusive) rocks are formed from magma that hardened underground. Such rocks are then later exposed after many thousands of years of erosion. The word *plutonic* comes from Pluto, the Greek god of the Underworld.

Igneous rocks come in many colors and varieties, resulting from differences in the kinds, sizes, amounts, and arrangements of minerals they contain. These differences are used by scientists in identifying, naming, and classifying igneous rocks.

Texture and Mineral Size

The texture of an igneous rock is determined by the amount and crystal size of the minerals it contains. Both these mineral properties depend on how quickly the magma cools and becomes solid rock. The deeper the magma lies below the earth's surface, the more slowly it cools and hardens—and the more time minerals have to crystallize and grow. In contrast, magma closer to the surface cools more rapidly, and fewer minerals can crystallize or grow before the magma becomes completely solid. When magma is erupted at the surface as lava, the cooling is extremely fast. In the case of explosive eruptions, cooling is virtually instantaneous, because the tiny bits of lava rapidly lose heat to the cold air. The liquid part of the lava is quickly chilled to form volcanic "glass." Chemists define *glass*—volcanic or man-made—as a noncrystalline material formed by the sudden cooling of silicate liquid.

Because they are formed by slow cooling and complete crystallization of magma, plutonic rocks are composed entirely of minerals with little or no glass. In comparison, the rapidly cooled volcanic rocks contain few minerals and much more glass. Some volcanic rocks are composed almost entirely of glass. However, even the glassiest volcanic rocks contain a few tiny crystals.

The crystal size of minerals (grain size) in an igneous rock is also controlled by how fast magma cools and hardens. Slow cooling of magma allows the formation of large crystals, and rapid cooling results in smaller crystals. Because crystal size varies for the minerals in a rock, scientists use an average grain size in describing the texture of an igneous rock. The term *aphanitic* (from the Greek word for "invisible") refers to a rock whose minerals are difficult or impossible to see without the aid of a magnifying lens. An aphanitic rock has an average grain size of less than .02 inch (.5 mm). Rocks with a larger average grain size are said to have *granular* texture; in such rocks the minerals can be seen easily with the naked eye. Granular igneous rocks are subdivided according to their average grain size: fine-grained (less than .04 inch

[1mm]),medium-grained (.04–.2 inch [1–5 mm]), coarse-grained (larger than .2 inch [5 mm]), and very coarse-grained (larger than 1.2 inches [30 mm]).

Some igneous rocks contain certain minerals which are conspicuously larger than the others. These rocks are called porphyries, and their texture is called porphyritic. Such textures are formed when there is a sudden change in the rate of cooling before the magma hardens completely. This can happen if magma, which has been cooling slowly deep in the earth, is erupted to the surface and cooled quickly. A similar texture results if the originally deep, slowly cooling magma rises quickly to a higher, colder level in the crust, even if it does not break the surface.

A rock composed of fragmental volcanic products—shattered rock fragments, broken crystals, and bits of volcanic glass—is described as having a pyroclastic texture. A plutonic rock would never have a pyroclastic texture. Many volcanic rocks have abundant gas-bubble holes, resulting in a so-called vesicular texture. Plutonic rocks do not exhibit a vesicular texture. In general, textures of plutonic rocks are relatively simple, whereas the textures of volcanic rocks, especially the pyroclastic varieties, can be more complicated.

Kinds of Minerals and Rock Color

Most igneous rocks, whether plutonic or volcanic, are composed mainly of silicate minerals. The basic building block of rock-forming minerals is the so-called silicate tetrahedron. This is a pyramid-shaped cluster of four oxygen ions around one silicon ion. Different ways of stacking these silicate tetrahedra and combining them with other chemical elements result in the formation of the many different kinds of silicate minerals. The common minerals in igneous rocks contain mostly oxygen (O) and silicon (Si). Other important chemical elements include aluminum (Al), iron (Fe), magnesium (Mg), calcium (Ca), sodium (Na), potassium (K), titanium (Ti), and manganese (Mn). Although they both have the same kinds of silicate minerals, plutonic rocks as compared with volcanic rocks have more, and larger, crystals and little or no glass. The silicate minerals listed below make up the common igneous rocks.

Light-colored Minerals

Quartz—a six-sided crystal composed only of oxygen and silicon.

Alkali feldspar—a boxlike crystal containing aluminum, potassium, sodium and a small amount of calcium in addition to oxygen and silicon.

Plagioclase feldspar—generally occurs as well-formed, brick-shaped crystals; contains much more calcium and sodium than alkali feldspar.

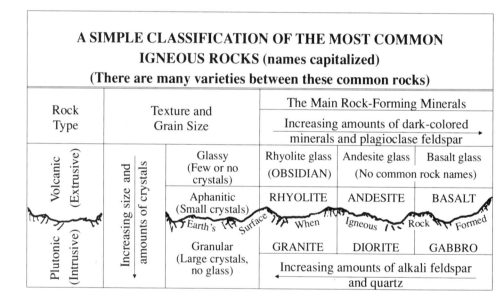

A SIMPLE CLASSIFICATION OF THE MOST COMMON IGNEOUS ROCKS (names capitalized)
(There are many varieties between these common rocks)

Rock Type	Texture and Grain Size		The Main Rock-Forming Minerals		
			Increasing amounts of dark-colored minerals and plagioclase feldspar →		
Volcanic (Extrusive)	Increasing size and amounts of crystals	Glassy (Few or no crystals)	Rhyolite glass (OBSIDIAN)	Andesite glass (No common rock names)	Basalt glass
		Aphanitic (Small crystals)	RHYOLITE	ANDESITE	BASALT
Plutonic (Intrusive)		Granular (Large crystals, no glass)	GRANITE	DIORITE	GABBRO
			← Increasing amounts of alkali feldspar and quartz		

Dark-colored Minerals
(also called "ferromagnesian" minerals)

Olivine—usually occurs as well-shaped yellowish-green to dark green crystals; in addition to the silicate tetrahedra, contains mostly iron and magnesium.

Pyroxene—usually occurs as stubby, eight-sided crystals dark green to black in color; contains mainly calcium, iron, magnesium, and aluminum in addition to the silicate tetrahedra.

Amphibole—commonly occurs as six-sided crystals, somewhat longer than the pyroxene minerals; similar to pyroxene in composition but contains a little water locked in its crystal structure.

Mica—usually six-sided flake-like crystals; in addition to silicate contains much potassium, along with variable amounts of magnesium, iron, and aluminum. The more iron and magnesium it contains, the darker the crystal. Like the amphiboles, the mica minerals also contain a small amount of water in their crystal structure.

In addition to the minerals listed above, glass can make up a large portion of a volcanic rock. Various other minerals can occur in small amounts in the common igneous rocks, but they are not usually important in identifying or naming the rock.

Igneous rocks are identified as either plutonic or volcanic, and are named according to their texture, mineral content, grain size, and general color. For example, gabbro is the plutonic equivalent of the volcanic rock basalt, because they have the same magma composition and the same kinds of minerals. But they differ greatly in amount of minerals and in texture. Gabbro has a granular texture, whereas basalt can have an aphanitic, glassy, or pyroclastic texture. Similarly, the volcanic rock rhyolite corresponds to the plutonic rock granite: Both are derived from the same magma, which forms rhyolite if it is erupted as lava, and granite if it slowly crystallizes and hardens deep in the earth.

Scientists have recognized and named well over a hundred types of igneous rocks. However, the most common rock types, and the intermediate varieties between them, would make up more than 95 percent of the igneous rocks found on the earth's suface.

With rare exceptions, the color of an igneous rock is determined by the amount of dark-colored (or ferromagnesian) minerals it contains. In turn, the amount of dark-colored minerals that can crystallize from a magma depends on the amount of iron, magnesium, and calcium it contains. Rhyolite and dacite (an intermediate variety between rhyolite and andesite) are light-colored volcanic rocks, because the lavas from which they are formed contain little iron, magnesium, and calcium as compared with silicon, aluminum, sodium, and potassium. Basalt is a very dark-colored volcanic rock because of its much higher content of iron, magnesium, and calcium, and lower amounts of silicon, aluminum, sodium, and potassium. Obsidian is an exception. Being very high in silicon and potassium, and very low in iron and magnesium, it should be light-colored, but instead it is always

The difference in color and texture between Mount St. Helens dacite (at right) and Hawaiian basalt results from differences in magma composition and rate of cooling.

44

shiny black or dark brown in color. Obsidian appears dark because its transparent glass is tinted by many thousands of tiny ferromagnesian minerals, even though the amount of iron and magnesium is very small. Thus, even though obsidian, rhyolite, and the plutonic rock granite all have about the same chemical composition, they look very different in texture and color.

Forms of Plutonic Rocks

Plutonic rocks can occur in many forms. These plutonic rocks, though originally formed beneath the earth's surface, are eventually exposed by erosion, allowing scientists to map and study them. At some places, where erosion has cut very deep, volcanic rocks on a mountaintop or ridge can be followed downward into plutonic rocks exposed in a canyon. This would give direct proof that the volcanic and plutonic rocks were formed from the same magma. More often, however, any connection between volcanic and plutonic rocks in the same region must be shown by indirect evidence. Careful mapping and laboratory studies of the rocks are needed to demonstrate that they are related members of the same magma family.

Much of the earth's plutonic rock is exposed as huge masses called *batholiths*, some of which cover many thousands of square miles. Similar but smaller occurrences, covering less than 40 square miles (100 sq. km) are called *stocks*. One of the largest and best studied batholiths is the Sierra Nevada batholith of California. Part of it is spectacularly exposed in the magnificent granite cliffs of Yosemite National Park. Around the edges of batholiths or stocks, especially in their upper part (called the roof), blocks of nonplutonic rocks are commonly found. These blocks, called inclusions, are pieces of the old solid surrounding rock that were caught up in the rising magma before it hardened and crystallized completely. Such inclusions can range in size from small fragments of one inch (2.5 cm) or less to huge blocks tens of miles across.

A pluton or intrusion means any mass of plutonic rock, regardless of its size or shape. However, because many small intrusions have quite regular or distinctive shapes, they have been given special names. The most common of these special types of intrusions include: sill, dike, laccolith, lopolith, and volcanic neck or plug. The distinctions between these intrusions are based on their occurrence and shape. The most important consideration is how the original magma was injected (or intruded) into the surrounding strata. Sills and dikes are long but thin, slab-like intrusions, rarely more than tens of yards thick. Sills are sandwiched between—and thus parallel to—the strata of surrounding rock. Dikes cut across the surrounding strata. Laccoliths and lopoliths are special cases of sills. They are not common.

Volcanic necks or plugs are the preserved fossil remains of the magma left in the bowels of a volcano, exposed by erosion long after it became inactive. Although they are not very common, volcanic necks or plugs have great geologic significance. Not only are they the

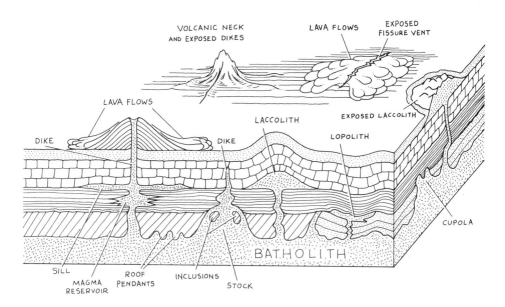

Simplified diagram illustrating the various forms of plutonic or intrusive rocks. (Modified from Emmons et al., 1960).

connecting link between plutonic and volcanic rocks, but sometimes they provide the only remaining direct evidence of the existence of an ancient volcano in a deeply eroded region. Ship Rock in New Mexico is an exceptionally well-exposed example of a volcanic neck.

Half Dome, Yosemite National Park, California, is an excellent example of plutonic rocks of a large batholith exposed by deep erosion.

6

Volcanoes, Volcanology, and People

The oldest rocks on earth are igneous rocks that formed about 3.8 billion years ago. Volcanism almost certainly occurred even earlier, but any evidence of it has been erased by later volcanism, mountain building, metamorphism, weathering, and erosion. Clearly, volcanoes have played an important role since the earliest history of our planet. Since the formation of the earth 4.6 billion years ago, countless eruptions have created the volcanic mountains, islands, plateaus, and plains that make up much of the present earth's surface.

Volcanology

Geology is the science of the earth. *Volcanology* is the branch of geology that deals with volcanoes, especially those that are active or might become active. Scientists who study the earth are called geologists or geoscientists, and those who specialize in volcanology are called volcanologists.

Volcanology is a young science. Volcanic eruptions only began to be studied seriously in the late nineteenth century, after the catastrophic 1883 eruption of Krakatau Volcano (Indonesia) that killed 36,000 people. However, volcanology really did not become a modern science until the early 20th century, when scientific volcano observatories were founded in Japan, in Hawaii, and in a few other countries with active volcanoes. The establishment of such observatories was prompted by three volcanic disasters in 1902 in the Central America-Caribbean region.

Predicting Eruptions

The main function of a volcano observatory is volcano monitoring—the round-the-clock observation and measurement of changes in the behavior of an active volcano between and during eruptions. Some changes are visible, such as the appearance of new ground cracks, widening or narrowing of old cracks, increased or decreased steaming at known vents, formation of new vents, unusual dying of plants, and so forth. Other changes too, though invisible, can be measured by volcanologists. Important measurable changes include: the shape of the volcano's surface, the ground temperature, the composition of waters, gases, and lava from vents, and the earth's gravity, magnetic, and electrical fields.

Changes at a volcano are caused by magma moving toward the surface to erupt. Underground magma movement can be studied by continuously measuring a volcano's earthquake (or *seismic*) activity. As the magma moves underground, it must make room for itself by breaking or pushing aside the surrounding solid rocks. Breaking of solid rock causes sharp ground motions (earthquakes), which send seismic shock waves through the volcano. These waves are picked up by instruments called seismometers, converted into electronic signals, and sent by radio to be recorded on other instruments called seismographs at the volcano observatory. The seismograph data are then analyzed to determine the time, location, and size (magnitude) of the earthquakes. Magma movement may also produce a continuous vibration of the ground called harmonic tremor or volcanic tremor. Such seismic tremor often occurs before and during eruptions.

Before an eruption, a volcano may puff up because of the forces exerted by rising magma. This process is like the stretching of a balloon being inflated. The puffing up or inflation of the volcano results in the steepening of its slope (tilt), changes in vertical and horizontal distances between reference points (benchmarks) on the volcano surface, and an increase in the number of earthquakes. Most changes in volcano shape are very small but can be measured precisely by scientific instruments and field-surveying techniques. For example, volcanologists use computerized electronic-laser beam equipment to measure changes in horizontal distance. Changes as small as a few parts per million (ppm) can be measured routinely. The change in tilt of a mile-long wooden plank when a coin is placed under one end is equal to one ppm!

For Hawaiian volcanoes, pre-eruption inflation is generally slow and gradual, lasting from weeks to years. But once the eruption starts,

PRE-ERUPTION INFLATION

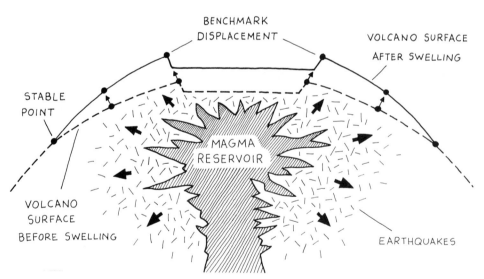

Simplified cross section of a volcano showing the changes in volcano shape that can be precisely measured when magma enters its reservoir.

an abrupt sagging (or deflation) occurs as pressure on the magma reservoir is relieved. During deflation, the measured changes in volcano shape are opposite to those during inflation. Volcano tilt and the horizontal and vertical distances all decrease, and the number of earthquakes drops to a normal level.

Volcano monitoring provides some of the information needed to predict a volcano's future behavior. But it is also necessary to learn about the volcano's past behavior and eruptions. Because the historical information for most volcanoes covers only a short time span, geologic studies of the prehistoric volcanic rocks are needed to answer important questions such as: How often did the volcano erupt in the past? What type of eruption—explosive or nonexplosive? How large? What was the average quiet time between eruptions? Without a good knowledge of past eruptive behavior over a long time period, and without learning the present behavior from volcano monitoring, volcanologists cannot make reliable forecasts of future eruptions.

Volcanologist making measurement of horizontal-distance change in volcano monitoring.

Only a few of the world's active volcanoes have been well studied, and even fewer are being monitored by modern volcano observatories. For most volcanoes, scientists still cannot predict eruptions, especially the large explosive ones. But progress is being made. For example, for Kilauea and Mauna Loa, volcanologists at the Hawaiian Volcano Observatory can tell when an eruption is building up. Sometimes they even know where the next eruption is likely to occur, but they cannot predict exactly when it will begin. Since June 1980, volcanologists at the Cascades Volcano Observatory (Vancouver, Washington) have successfully predicted nearly all the dome-building eruptions at Mount St. Helens. These predictions were announced publicly from several hours to a few weeks in advance—quite an accomplishment for a young science.

KILAUEA VOLCANO, HAWAII

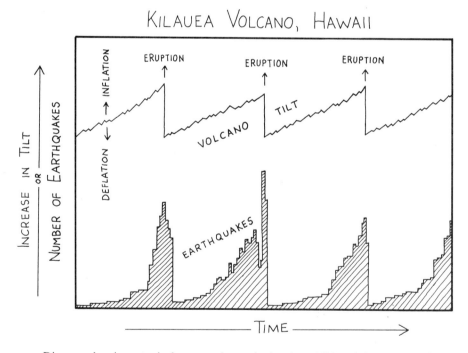

Diagram showing a typical pattern change in the slope (tilt) and the number of earthquakes seen during eruptions of Kilauea Volcano, Hawaii.

Evaluating Volcanic Hazards

While volcano monitoring may help predict eruptions, this by itself is not enough. The people responsible for handling a potential volcanic emergency must also have information on what to expect from a volcano should it erupt. They need to have answers to questions such as: What are the likely volcanic hazards to be expected? How serious or severe might these be? What areas are likely to be the hardest hit, and by which type or types of hazard?

In trying to answer these and similar questions, scientists make an evaluation of volcanic hazards based on all available geologic information. The more geologic data that goes into the evaluation, the more detailed, reliable, and useful that evaluation will be. The basic information needed for a useful hazards evaluation includes, 1) the kinds, sizes, and impacts of previous eruptions, historical and prehistoric; 2) a complete record of the dates and durations of historical eruptions; and 3) the geologic ages of the prehistoric eruptions. Such information can only be obtained by years of geologic mapping and laboratory studies, including the determination, by means of modern dating techniques, of the ages of the volcanic rocks.

A detailed geologic map of volcanic deposits of known ages allows the scientists to answer questions such as: What parts of the volcano have been hit in the past by pyroclastic flows? How many times? How thick were these flows? How far did they travel from the volcano? Similar questions can be answered for ash falls, mudflows, lava flows, and other hazards. Perhaps the most useful part of any complete evaluation is a map of volcanic-hazards zonation. Such a map shows not only the most hazardous zones, but also the least hazardous. Knowing both is important in planning evacuation routes, locating emergency shelters, and deciding on other public-safety measures.

A volcanic-hazards evaluation should be done for every active or potentially active volcano long before any volcanic emergency occurs. Emergency plans will then be immediately available when needed, instead of being called for at the last minute. A volcanic emergency

53

can develop too fast and leave little or no time for planning and carrying out emergency measures. Also, if government officials had good information about volcanic hazards well in advance, they could make better decisions about how best to use land in volcanic areas. For example, if studies show that a certain area is very likely to be in the path of pyroclastic flows from future eruptions, it would make good sense not to put high-rise buildings or a housing subdivision there!

Volcanism and Civilizations

Over geologic time—thousands to millions of years—volcanic rocks were broken down, weathered, and eroded to form some of the most

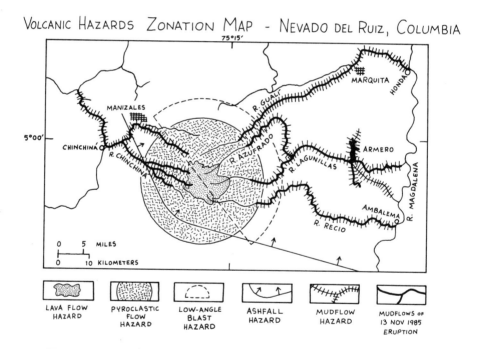

Volcanic Hazards Zonation Map - Nevado del Ruiz, Columbia

The volcanic-hazards zonation map that was prepared for Nevado del Ruiz, Colombia, in early October 1985, a month before its deadly eruption on November 13. The city of Armero was identified as being in the hazard zone danger from mudflows, but more than 22,000 people died because proper emergency action was not taken in time.

fertile soils and scenic landscapes on earth. It is no accident that many civilizations developed in volcanic areas, as those of the Greeks and Romans did in the Mediterranean region. As these civilizations grew, however, people had to live closer and closer to the active volcanoes. But despite occasional volcanic disasters, like the eruption of Vesuvius which destroyed Pompeii in A.D. 79, the populations around volcanoes continued to expand.

Some eruptions have been blamed for possibly changing the course of some ancient civilizations. For example, a tremendous eruption of Merapi Volcano is supposed to have devastated central Java, Indonesia, in the year A.D. 1006. Some claim that this catastrophic eruption caused, or at least quickened, the fall of the Hindu culture on Java in the eleventh century and the later rise of the Muslim culture in its place. This eruption supposedly terrified the then predominantly Hindu people living on Java and prompted them to resettle on the neighboring island of Bali. However, this idea is not accepted by all scientists. There is even some doubt whether a large eruption occurred at all at Merapi in A.D. 1006!

In about 1600 B.C., Santorini Volcano (on the Greek island of Thera in the Aegean Sea) erupted violently. A royal city of the Minoan civilization on Thera was destroyed. This catastrophic eruption of Santorini has been associated with the sudden decline of the centuries-old Minoan civilization in the Aegean region. Recently, however, some archaeologists have found new evidence that suggests that the Minoan culture survived the Santorini eruption and later collapsed for other, as yet unknown, reasons. Some scholars even speculate that the Santorini eruption gave rise to the ancient Egyptian and Greek legends about Atlantis, a powerful island empire that suddenly and mysteriously vanished. Most scientists, however, do not believe that the mythical Lost City of Atlantis was in any way connected with Santorini.

What about the impact of future huge eruptions on modern civilizations? Because the world is much more crowded now than during ancient times, eruptions threaten many more people—about 360 million in fact. Another eruption of Vesuvius as big as the one that destroyed Pompeii in A.D. 79 would be much more deadly today. The problem of densely populated areas near active volcanoes is even more serious for several Pacific Rim countries, among them Indonesia, Japan, and the Philippines. Unfortunately, most of the world's dangerous volcanoes are located in countries that do not have enough money or trained scientists to do the studies needed to reduce volcanic hazards.

Since the beginnings of civilization, no explosive eruption the size of those that produced the Yellowstone caldera has occurred. Such caldera-forming eruptions can be many hundred times larger than the largest known historic eruption (Tambora in 1815). Luckily, these

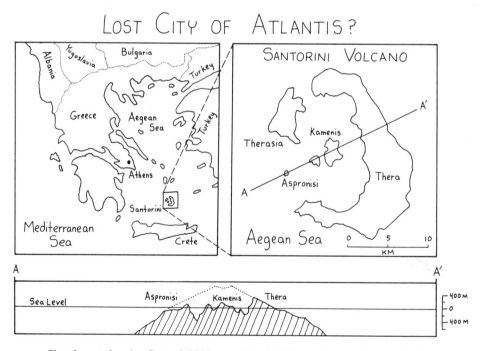

Sketch map showing Santorini Volcano, Island of Thera, Greece. Some scientists speculate that Santorini might be the site of the legendary "Lost City of Atlantis," presumably destroyed during a huge explosive eruption and caldera collapse there around 1600 B.C.

huge eruptions occur very infrequently, tens to hundreds of thousands of years apart. The more we understand about these gigantic eruptions in the geologic past, the better chance we have to prepare for the next Big One. It is not a question of if, but of when.

Short-term Hazards versus Long-term Benefits of Volcanism

Volcanic eruptions are disastrous in the short run, but beneficial over the long run. In addition to the volcanic gift of fertile, life-supporting soils, volcanism also helps mankind in other ways. Many volcanic regions are attractive places in which to live or vacation. Without volcanic activity, the beautiful chain of tropical islands that make up Hawaii, our fiftieth state, would not exist. The spectacular, forested mountains of the U.S. Pacific Northwest owe their beauty to their volcanic origin. Without volcanism, the hot springs, geysers, and other natural wonders of Yellowstone National Park would not have been formed.

Many of the world's richest ore deposits (copper, iron, gold, silver, lead, zinc, sulfur, etc.) are found in or near the roots of extinct volcanoes. This is because the deep, hot roots of volcanoes furnish heat to move and circulate ore-bearing fluids. When the temperature and pressure conditions are right, the ore metals in these fluids crystallize to form rich mineral veins. Volcanic rocks themselves provide construction or roadbuilding materials, abrasive and cleaning agents, and raw ingredients for many manufacturing uses.

Another benefit is the internal (geothermal) heat of active, or inactive but still hot, volcanic systems. Geothermal energy can be used to warm homes and greenhouses, generate electricity, and provide heat for some industrial processes. For example, about 80 percent of the homes in Iceland are heated by piping in and circulating naturally hot volcanic waters. In the United States, if all of the usable geothermal heat were developed, it could meet about 10 percent of our nation's energy needs. In certain areas, geothermal energy alone is more than enough. For example, the Geysers geothermal field of California—the largest in the world—produces enough electrical energy to supply the entire city of San Francisco. A third of all the power used in the country of El Salvador, Central America, is geothermal. In recent years, the geothermal energy of Kilauea Volcano, Hawaii, has been tapped by a small demonstration power plant which produces about three megawatts of electricity.

In summary, volcanoes are both a curse and a blessing. On a human time scale, eruptions can be hazardous. But over geologic time, volcanoes have benefited, and will continue to benefit, mankind. There is no scientific reason to think that volcanoes will be less active in the future than now. So, as world population grows, the problems posed by volcanic hazards will become worse. We must continue to improve our ability to predict eruptions and to make better evaluations of volcanic hazards. The volcanologists' job is to provide the best possible scientific information on high-risk volcanoes. This information can then be used by government officials for making sound decisions on land-use planning and, if necessary, for taking emergency measures, when future volcanic eruptions occur.

Glossary

aphanitic—Refers to a rock whose minerals are difficult or impossible to see without the aid of a magnifying lens; compare with granular.

ash fall—Airborne fine-grained fragmental lava that is carried and deposited downwind from explosive eruptions.

batholith—Large body of plutonic rock that can cover many thousands of square miles; compare with stock.

caldera—A huge circular crater, caused when a volcano collapses into the magma chamber below.

cinder—Fragmental lava, of irregular shape and greater than 2 1/2 inch (64 mm) in size, erupted in liquid or solid condition; also commonly called scoria

convergent boundary—A zone where neighboring tectonic plates collide and one of the plates is commonly dragged down (subducted) beneath the other; also called subduction zone.

crater—A hole or small depression (smaller than a caldera) on a volcano; commonly also a vent for eruptions.

crust—The outermost layer of the earth, it lies above the mantle and averages about 22 miles (35 km) thick; it is thickest beneath high continental mountain ranges.

crystal—Another name for mineral; commonly used for minerals that have well-defined shape and faces.

crystalline—Refers to rocks or materials that are mostly composed of minerals rather than glass.

dike—A slablike body of igneous rock (intrusion) that cuts across the layers of older rocks into which it has intruded; compare with sill.

earthquake—A sudden motion or trembling of the earth caused by the abrupt release of energy associated with ground rupture and faulting.

earthquake wave—A general term for various kinds of vibrational waves produced by an earthquake.

extrusive—Refers to products and processes involving nonexplosive eruptive activity; often used as a synonym for volcanic.

geology—The study of the earth in the broadest sense; there are many specialized branches of geology, including volcanology and seismology.

geothermal—Refers to the natural internal heat of the earth, most abundantly found in volcanic areas.

glass—A noncrystalline material—either of man-made or volcanic origin—formed by the sudden cooling of silicate liquid.

grain size—Refers to the general or average size of minerals that make up a rock.

granular—Refers to a rock whose minerals can easily be seen with the naked eye; compare with aphanitic.

hot spot—A term used in the plate-tectonics theory that refers to a place where a fixed heat source in the mantle that partly melts the solid rock of the overriding plate to produce magma, which is later erupted to form new volcanoes.

igneous—A term applied to all processes and products involving magma; synonymous with magmatic.

lahar—An Indonesian term commonly used for volcanic mudflows.

lava—Refers to magma (molten rock) that has reached the surface: applied also to the rock formed after it cools and hardens.

lava flow—A stream of molten rock (lava) flowing from a volcanic vent; also used for the rocks formed after the lava cools and hardens.

magma—Refers to the underground molten rock as well as the gases dissolved in it; called lava once it has erupted to surface.

mantle—The layer of the earth that lies beneath the crust and above the core, extending to a depth of about 2160 miles (3480 km).

mineral—A substance with an internal structure determined by a regular arrangement of atoms; generally has a chemical composition that is fixed or varies within certain limits.

phreatic—Refers to volcanic rock; synonymous with steamblast.

plate tectonics—A widely accepted theory that the outer part of the earth is broken into a dozen or so rigid slabs (or plates) that move slowly relative to one another.

pluton—An igneous body formed by the cooling and hardening of magma that never reached the surface; also called an intrusion.

plutonic—Refers to any product or process involving magma beneath the earth's surface; synonymous with intrusive, magmatic.

porphyritic—A texture of an igneous rock that contains some minerals conspicuously larger than the others.

pumice—Glassy fragmental lava erupted during violently explosive eruptions; because it contains many vesicles (gas-bubble holes),pumice is very light.

pyroclastic—A term meaning "fire-broken" in Greek, refers to any shattered volcanic material, either of old solid rock or new molten lava, thrown out during explosive eruptions.

Ring of Fire—The name given to the notorious zone of frequent earthquakes and eruptions fringing the circum-Pacific region.

scoria—See cinder.

seismic—Refers to earthquake processes.

seismic wave—See earthquake wave.

seismology—A special branch of geology that studies earthquakes.

silicate—A chemical compound formed primarily by the combination of the elements silicon and oxygen; silicate minerals form the overwhelming bulk of the earth's mantle and crust.

stock—A body of plutonic rock similar to a batholith but smaller, covering less than 40 square miles (100 sq km).

subduction zone—See convergent boundary.

texture—the appearance of a rock determined by the kind, amount, and size of the minerals it contains.

transform fault—The boundary along which one tectonic plates slides horizontally past another.

vent—The opening through which volcanic materials—whether solid, molten, or gaseous—are erupted.

viscosity—A measure of resistance to flow in a liquid; for example, water has a low viscosity while tar has a high viscosity.

volcanic—Refers to products or processes involving volcanoes and eruptions.

volcanic ash—Fragmental lava finer in size than 1/10 inch (2 mm).

volcanic block—Fragmental lava, coarser than about 1 1/2 inch (64 mm) and of angular shape, erupted in completely solid condition; compare with volcanic bomb.

volcanic bomb—Fragmental lava, coarser than about 2 1/2 inch (64 mm) and having round to slightly angular shape, erupted in a semisolid condition; compare with volcanic block.

volcanic neck—The fossil remains of magma left in the throat of a volcano, exposed by erosion long after it became inactive; also called volcanic plug.

volcano—A mountain or hill formed of molten or solid material erupted from openings in the earth's surface. Also refers to the opening (or vent) itself.

volcanology—A special branch of geology that specializes in the study of volcanoes, especially those that are active or might become active.

volcano observatory—A research facility that makes observations and measurements of active volcanoes before, during, and after eruptions.

Further Reading

Aylesworth, Thomas G. *Moving Continents: Our Changing Earth.* Hillside, NJ: Enslow Publishers, Inc., 1990.

Barker, Daniel S. *Igneous Rocks.* Englewood Cliffs, NJ: Prentice-Hall, Inc., 1983.

Blong, Russell J. *Volcanic Hazards: A Sourcebook on the Effects of Eruptions.* Sydney, Australia: Academic Press, 1984.

Chester, D.K., A.M. Duncan, J.E. Guest, and C.R.J. Kilburn. *Mount Etna: The Anatomy of a Volcano.* Palo Alto, CA: Stanford University Press, 1985.

Cole, Joanna. *Magic Schoolbus: Inside the Earth.* New York: Scholastic, 1987.

Decker, Robert W. and Barbara Decker. *Volcanoes.* 2nd ed. New York: W.H. Freeman and Company, 1989.

Harris, Stephen L. *Fire Mountains of the West: The Cascade and Mono Lake Volcanoes.* Missoula, Mont.: Mountain Press Publishing Co., 1988.

Hoffer, William. *Volcano: The Search for Vesuvius.* New York: Summit Books, 1982.

Lauber, Patrica. *Volcano: The Eruption & Healing of Mount St. Helens.* New York: Bradbury Press, 1986.

Montgomery, Carla W. *Physical Geology.* Wm C. Brown Publishers: Dubuque, Iowa, 1987. [Good chapter on igneous activity and igneous rocks]

Planet Earth: Volcano. Alexandria, VA: Time-Life Books, 1982.

Poynter, Margaret. *Earthquakes: Looking for Answers.* Hillside, NJ: Enslow Publishers, 1990.

Ritchie, David. *The Ring of Fire.* New York: The New American Library, Inc., 1981.

Simkin, Tom and Richard S. Fiske. *Krakatau 1883: The Volcanic Eruption and Its Effects.* Washington, DC: Smithsonian Institution Press, 1983.

Simkin, Tom, Robert I. Tilling, James N. Taggart, William J. Jones, and Henry Spall. *This Dynamic Planet: World Map of Volcanoes, Earthquakes, and Plate Tectonics.* Reston, VA: U.S. Geological Survey, 1987.

Tilling, Robert I., Lyn Topinka, and Donald A. Swanson. *Eruptions of Mount St. Helens: Past, Present, and Future.* Reston, VA: U.S. Geological Survey, 1990 (revised edition).

Tilling Robert I. *Eruptions of Hawaiian Volcanoes: Past, Present, and Future.* Reston, VA: U.S.Geological Survey, 1987.

Wenkam, Robert. *The Edge Of Fire: Volcano and Earthquake Country in Western North America and Hawaii.* San Francisco, CA: Sierra Club Books, 1987.

INDEX